Waterworks

Water Play Activities for Children Aged 1-6

Jeanne C. James
Randy F. Granovetter
Illustrations by: Sheila Krill

Published by:
Kaplan Press
1320 Lewisville-Clemmons Road
Lewisville, N.C. 27023

Dedication

For Jason, Brett and Jo Anna who taught us how to frolic and learn through water play.

For Anne Sanford who brought us together and taught us the true meaning of dedication.

Contents

Introduction

We wrote this book because water play provides both fun and valuable learning opportunities for young children. Water play can provide calming activities for any child. Water play activities can reinforce the principles of scientific experimentation and mathematical reasoning. They integrate cognitive, fine motor and gross motor skills. We encourage all teachers, caregivers and parents to provide water play opportunities for the young children in their care.

Instructor's Guide

Before discussing the format of the book, we ask that you be certain to read the Safety Tips which follow the Instructor's Guide.

This book is organized into three sections. Each section consists of activities for a particular age group. These are:

        Activities for ones and twos;
        Activities for threes and fours; and
        Activities for fives and sixes.

Within each section the activities are divided into five types of activities. The types of activities are:

Activities to be done with a small amount of water by one or more children. These include pouring and mixing, pump art, spray art, sink and float, and blowing bubbles. The logo for these activities is:

Activities for a water table which can be done by one or more children. The logo for these activities is: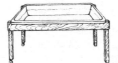

Activities which can be done in a sprinkler. The sprinkler should not move, but consist of a head which sits in or on the ground and sprays water in a circle around it like the logo which is:

Activities for a small pool which are preparatory for getting into a large pool. The logo is:

Activities for a large pool to acquaint the children with the pool and prepare them for swimming. The logo is:

Familiarity with the logos will help you to locate particular activities quickly.

Safety Tips

Always have an adult with the child or children in any water play situation. NEVER LEAVE A CHILD UNATTENDED.

Use only unbreakable materials for water play activities. NEVER USE GLASS, CERAMIC, PORCELAIN, POTTERY, CHINA OR OTHER BREAKABLE MATERIALS.

Always gather materials ahead of time, so you do not have to leave the child or children unattended.

Develop water play rules with the children. Discuss and generate a list that is appropriate to your situation and revise them as needed. For example, you will need one set of rules when the children are playing at the water table and other rules for the large pool.
> Some ideas for rules include:
> > No splashing.
> > Keep the water in the containers.
> > Mop up spills immediately.
> > No running in the pool area.

Supervise children carefully and remind them of the water play rules.

Explain and discuss water play rules with the children.

Materials

Many of the materials used in water play are found objects such as empty pump, spray and squeeze dishwasher detergent bottles, measuring caps from coffee and drink mixes, styrofoam trays, styrofoam pieces, rocks, etc. In a day care or nursery setting parents can save these materials for use by the children. Clean materials of this type thoroughly.

Other materials are readily available from local sources. These include construction paper, dishwashing liquid, scissors, tape, food coloring, sponges, ping-pong balls, ribbon, markers, straws, plastic measuring cups, styrofoam cups, dishpans, sugar, balls, sprinklers, etc. Check local discount grocery and hardware stores.

Other materials will have to be purchased from specialized sources such as school supply catalogs. This includes materials such as wind up toys, water wheels, plastic tubing, clear basters, water tables, small pools, etc.

Always gather all the materials for an activity before you begin so that you will not have to leave the children alone to get them.

Infant Waterplay Activities

There are many natural teaching opportunities that occur in your daily routine with your baby. One special time is bathtime. Listed below are some suggested infant stimulation activities to do with your baby during bathtime.

Ages 0-6 months

During 0-6 months the baby is developing and sharpening his/her senses of taste, touch, smell, hearing, vision, and motor development. The baby cannot reply but he/she loves listening to your voice, looking at and later exploring the objects present in a given situation. Most babies are beginning to enjoy bathtime by the second month. They delight in kicking, splashing, and hitting the water. Your baby's motor development is increasing daily, and it is a good rule to have one hand on the baby at all times to avoid the danger of the child falling off of a changing table, bed, or slipping into the water.

FOLLOW THE DRIPS:

Have a plastic net bag or bucket filled with objects of different textures such as sponges and washcloths (terrycloth and velour).
Dip the sponge or washcloth in water. Hold the sponge or washcloth in the air and squeeze the water out of the washcloth onto the child's leg, making an arc of drips. Repeat the activity until the child is anticipating and watching the water dripping motion. When you go to fill the sponge or washcloth with water again, rub the cloth gently on the child's leg, arms, or stomach. Pair the actions with words such as, ''Ready, here come the drips!'' Pairing an action with a verbal cue, helps prepare the child that something is going to happen so that he/she will soon recognize that statement and situation with pleasant activity that follows and become excited anticipating the bathtime play.

WATERFALL:

Have a bucket or plastic bag with unbreakable cups (paper, plastic, or styrofoam), a baster, an eye dropper, and small measuring cups. Fill the cup, baster, eye dropper, and/or measuring cup up with water and pour it gently onto the child's leg. Repeat the activity so that the child will anticipate and watch the water pour out of the cups and stream out of the baster or eye dropper. Pair the actions with a verbal cue such as, ''Ready, here comes the waterfall!''

Ages 6-12 months

The 6-12 month old is sitting, crawling, and practicing motor movements in preparation for walking. He/she is constantly exploring his/her environment due to this new freedom of movement and constant development of keener eye-hand coordination. During this stage of development, the child is constantly picking up objects and exploring the different ways to use them. Bathtime is a favorite time of the day when pouring and dumping of liquids is allowed.

POURING AND DUMPING:

Provide a plastic container of different pouring objects of different sizes (unbreakable cups, measuring cups, spoons) and a plastic bucket. Show the child how to fill the cup and pour it into the bucket. When the child fills the bucket with water help him/her dump the bucket and begin filling again.

WATERWHEEL:

This is a good age to purchase or make a waterwheel (see Make a Rube Goldberg Activity). Show the child how to pour the water over the waterwheel to make the wheels move.

GO FISHING

Fill a tub with different colored stacking rings. Have the child reach for the floating rings, pick them up, and stack them on the ring. Repeat the activity, putting the rings in different places floating in the tub.

ACTIVITIES FOR
ONE AND TWO YEAR OLDS

Pouring and Mixing Back and Forth

Materials:

- Clear plastic cups
- Small pitchers
- Water
- Red and blue food coloring
- A pan to catch spills if indoors

Materials Preparation:

None

Activities:

The child:

- Pours water from a small pitcher into a cup.
- Pours the water back into the pitcher.
- Pours water from one cup to another.
  or
- Pours water into a cup in which the adult has placed a few drops of food coloring.
- Stirs the mixture with a spoon.

The adult:

- Shows the child how to pour.

Things To Think About

Basic Skill: Pouring from one cup to another.

Pump Art
Chalk Pictures

Materials:

- White Chalk
- Plastic pump bottle
- Water
- Small chalk board
- Towel or mat

Materials Preparation:

- Put a chalkboard outside on the ground or put a towel or mat under the chalkboard inside the room.
- Fill the plastic pump bottle with water.

Activity:

The child:

- Colors on the chalk board with white chalk.
- Presses the pump top and lets the water splatter over the chalkboard, making a water design on the chalkboard.

The Adult:

- Shows the child how to press the pump to get the water out of the plastic bottle.

Things To Think About

Basic Skill: Pushing a pump top.

Squeezing Activities
Fill 'Em! Squirt 'Em!

Materials:

- Dishpan-⅓ full of water
- Assorted squeeze bottles (small, clean, clear, plastic)
- Clear and/or translucent-plastic basters

Materials Preparation:

- Place the dishpan of water at standing height for the child

Activity:

The child:

- Unscrews the lid of the bottle.
- Fills the bottle by holding it underwater in the dishpan.
- Replaces the lid with adult help.
- Empties the bottle by squeezing the water back into the dishpan.
  or
- Places the tip of the baster underwater.
- Squeezes the bulb.
- Releases it to pull in the water.

The adult:

- Points out to the older child that air bubbles out of the bottle as water fills it.
- Helps the child to replace the lid.
- Discusses with the child when the bottle is full and empty.
- Points out to the child when the bottle is heavy and not heavy.

Things To Think About

Clear bottles allow the child to see the water fill them.
Small squeeze bottles are easy for the child to grasp.
Soft squeeze bottles are easy for the child to squeeze.

Basic Skill: Unscrews lid, fills and empties, stands alone well.

Spray Art
Picture Making

Materials:

- Squeeze plastic spray mist bottle (with trigger handle)
- Food coloring
- Large white butcher block paper
- Scissors
- Tape
- Large towel or plastic mat
- Water

Materials Preparation:

- Clean and rinse spray bottle.
- Spread towel or plastic mat on floor.
- Cut a large square of butcher block paper.
- Tape the paper to the towel or plastic mat.
- Fill the plastic spray bottle with water.
- Squeeze 6-10 drops of food coloring into the plastic spray bottle.
- Secure the top of the spray bottle.
- Prepare several plastic spray bottles of water with different colors of food coloring.

Activity:

The child:

- Shakes the plastic spray bottle.
- Stands next to the paper.
- Holds the spray bottle with both hands.
- Sprays the colored water onto the paper.

The adult:

- Removes the tape and hangs the picture.

Things To Think About

Basic Skill: Squeezing using a palmer grasp.

Blowing Bubbles
Blow Out For Bubbles

Materials:

- Cups (plastic, paper, styrofoam)
- Straws
- Water

Materials Preparation:

None

Activity:

The child:

- Puts the straw in the water.
- Blows into the straw to make the water bubble.

The adult:

- Practices blowing with the child until he/she understands how to blow through a straw.

Things To Think About

The young child may suck on the straw, so have him/her blow out of the straw onto his/her hand to feel the air coming out of the straw.

Basic Skill: Blowing out through a straw.

Going Fishing
Catch A Fish

Materials:

- Colored sponges
- Scissors
- Water table, water tub, sink, bucket, etc.
- Two buckets

Materials Preparation:

- Cut big and little fish out of sponges.
- Tape a "little fish" on the front of one bucket and a "big fish" on the front of the other bucket.
- Put the buckets next to the water table, bucket, sink, water tub, etc.

Activity:

The child:

- Upon the adult's direction, picks up a big fish or little fish.
- Squeezes the water out of the sponge fish.
- Puts the "big fish" in the bucket with a big fish on it and the "little fish" in the bucket with the little fish taped onto it.

The adult:

- Puts the fish shaped sponges in the water table, bucket, sink, water tub, etc.

Things To Think About

Basic Skill: Squeezing sponges.

Sink or Swim
Which Ones Float?

Materials:

- Water table ⅔'s full of water
- Assorted objects such as corks, bar soap, rocks, sponges, styrofoam, etc.
  CAUTION: no small objects!

Materials Preparation:

None

Activity:

The children:

- Put various objects into the water.
- Note which sink (fall to the bottom).
- Note which float (ride on top of the water).

The adult:

- Points out to the child which objects sink or float. Says "Look, that one is falling to the bottom" or "That one is riding on the top of the water" as appropriate.

Things To Think About

Basic Skill: Grasping objects with a palmer grasp.

CAUTION: no small objects!

16

Boats
Styrofoam Square Boats

Materials:

- Water table, water tub, sink, bucket, etc.
- Styrofoam square
- Straw
- Construction or typing paper
- Scissors
- Pencil (unsharpened)
- Markers or crayons
- One-hole punch

Materials Preparation:

- Cut out a styrofoam square.
- Prepare the sail for the child.
  Sail: Cut out a triangle from construction or white typing paper.
- Tape the paper triangle to the straw *or*
- Using the one-hole punch, punch a hole in the top and bottom of the paper triangle.
- Push the straw through the paper sail.

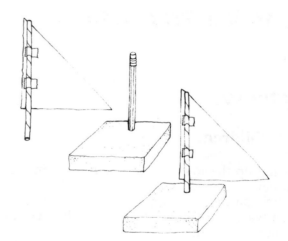

Activity:

The child:

- Pokes a hole in the styrofoam square with an unsharpened pencil.
- Places the straw sail into the hole in the styrofoam.
- Floats the styrofoam boats into the water.

The adult:

- Tapes the styrofoam square to the table so it will remain stationary for the child to make his/her pencil hole *or*
- Holds the styrofoam square so it will remain stationary for the child to make his/her pencil hole.

Things To Think About

Basic Skill: Pushing.

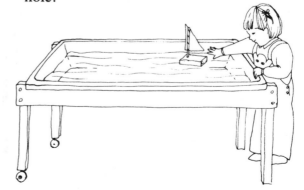

Races
Wind Them Up

Materials:

- Water table, water tub, sink, etc.
- Wind up bath toys
- Ribbon

Materials Preparation:

None

Activity:

The children:

- Wind up the toys and let them glide in the water.
- Wind up the toys, follow them as they glide in the water, rewinding them until they get to the other end of the table.

The adult:

- Shows each child how to wind up a small bathtub toy.
- Tapes or ties a ribbon down the middle of the water table, water tub, sink, bucket, etc.
- Has two children stand on either side of the ribbon.
- Clears the water table, water tub, etc. of everything but two wind up toys.
- Puts one toy on either side of the ribbon.

Things To Think About

Basic Skill: Winding a wind up toy.

Water Has Energy!
Water Turns a Wheel!

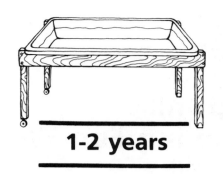

Materials:

- Water wheel
- Cups (plastic, paper, or styrofoam)
- Water table half full of water

Materials Preparation:

None

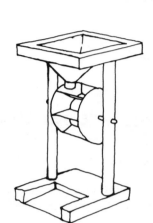

Activity:

The children:

- Fill cups with water from the water table.
- Pour the water onto the water wheel to make the wheel turn.

The adult:

- Shows the children how to make the water wheel turn.

Things To Think About

Basic Skill: Filling and emptying.

19

Sprinkler
Having Fun With Puddles

Materials:

• Sprinkler
• Sheet of plastic

Materials Preparation:

• Set sprinkler up in an area where it will make puddles.
• Set it up on plastic and make ridges to collect water if a natural puddle making area is not available.

Activity:

The children:

• Jump in the puddles as they collect.

The adult:

• Encourages the children to get under the sprinkler.

Things To Think About

Basic Skill: Jumping.

Sprinkler
Sprinkler Cup Filling

1-2 years

Materials:

Paper or styrofoam cups
Sprinkler

Materials Preparation:

None

Activity:

The children:

- Hold paper or styrofoam cups.
- Sit under the arch of the water from the sprinkler and fill the cups with water.
  or
- Hold paper or styrofoam cups.
- Walk through the sprinkler and fill the cups with water.

The adult:

- Shows the child how to catch water in a cup and fill it while sitting under the sprinkler.

Things To Think About

Basic Skill: Grasping cup, running.

Small Pool Ball Roll

Materials:

• Small pool
• Beach balls (different sizes)
• Rubber balls (different sizes)
• Tennis balls (different sizes)
• Plastic balls (different sizes)

Materials Preparation:

None

Activity:

The child:

• Rolls a beach ball back and forth to another child.
• Rolls different sizes and textures of balls back and forth.

The adult:

• Seats two children in the pool, facing each other.
• Gives the children a beach ball.
• Shows the children how to roll the ball back and forth in the water.

Things To Think About

Basic Skill: Rolling.

Small Pool
Ring Around the Rosie

1-2 years

Materials:

• Small pool filled with water

Materials Preparation:

None

Activity:

The children:

• Climb into the pool.
• Sing "Ring Around the Rosie."
• "All fall down" into the pool.

The adult:

• Teaches the children to sing "Ring Around the Rosie."

Things To Think About

Basic Skill: Follows directions in a song and cooperates in a group game.

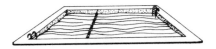

Large Pool
Pull and Twirl

Materials:

• Large pool
• One adult per child.

Materials Preparation:

None

Activity:

The children:

• Are held by the arms and twirled around in the water.

The adult:

• Holds the child by the arms.
• Twirls child around in the water.

Things To Think About

Basic Skill: Relaxes in the water.

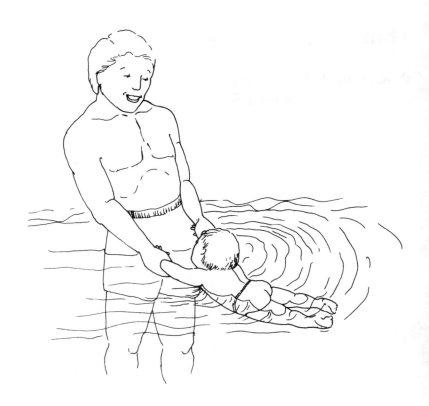

Large Pool Airplane

Materials:

• Large pool

Materials Preparation:

None

Activity:

The children:

• Hold arms straight out like the wings of an airplane.

The adult:

• Holds child under the stomach and chest.
• Circles each child in the water.

Things To Think About

Basic Skill: Relaxes muscles in water.

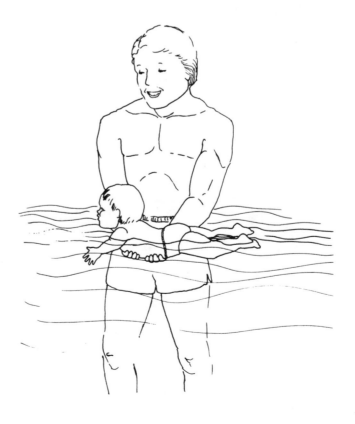

ACTIVITIES FOR
THREE AND FOUR YEAR OLDS

Pouring and Mixing Water Colors

3-4 years

Materials:

- Cups (plastic, paper, or styrofoam)
- Plastic pitchers
- Spoons
- Water
- Squeeze bottles of red, yellow and blue food coloring
- Paste food coloring in red, yellow and blue
- A shallow pan to catch spills if indoors
- Assorted construction paper

Materials Preparation:

- Place a dab of paste food coloring on some of the spoons.

Activities:

The child:

- Squeezes a couple of drops of one color of food coloring into each cup.
- Pours a small amount of water into each cup.
- Stirs the liquid to create evenly colored water.
- Blends the colors.
- Stirs the mixtures.
  (Start out with two colors like red and yellow which make orange).
  or
- Pours a small amount of water into each cup.
- Stirs with a spoon which has a dab of paste food color on it.

The adult:

- Helps the child identify the colors.

Things To Think About

Basic Skill: Stirring with a spoon.

Pump Art
Chalk Designs

Materials:

- Colored chalk
- Plastic pump bottles
- Water

Materials Preparation:

- Fill the plastic pump bottle with water.

Activity:

The child:

- Colors the sidewalk with colored chalk.
- Presses the pump top and lets the water splatter over the chalk on the sidewalk, making a water design on the chalk.

The adult:

- Shows the child how to press the pump to get the water out of the plastic bottle.

Things To Think About

Basic Skill: Pushing a pump top.

Squeezing Activities
Look at Me Squeeze

Materials:

- Dishpan-½ full of water
- Clear plastic squeeze bottles (large, medium and small, clean)
- Basters

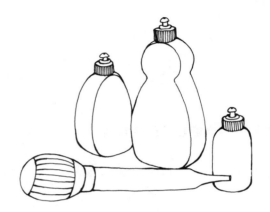

Materials Preparation:

None

Activity:

The child:

- Selects a large, medium or small squeeze bottle.
- Unscrews the lid.
- Holds it underwater in the dishpan until it is full.
- Replaces the lid.
- Empties the bottle by squeezing the water into the dishpan.
  or
- Puts the empty bottle in the water straight down.
- Notices when it does not fill.
  or
- Squeezes the baster while the tip is underwater.
- Observes the air bubbling out of it.
- Releases the baster bulb.
- Fills the baster with water.

The adult:

- Notices with the child that the bottle being emptied stays squeezed until it is turned so that air an get back into it.
- Points out to the child that air coming out of the baster makes bubbles.

Things To Think About

Basic Skill: Understands size differences.

Spray Art
Mural Making

Materials:

- Spray mist bottle with trigger handle
- Squeeze bottle of food coloring
- Large white butcher block paper or a white sheet
- Scissors
- Tape
- Large towel or plastic mat
- Water
- Table top

Materials Preparation:

- Spread towel or plastic mat on floor.
- Cut a large square of butcher block paper.
- Tape the paper or the sheet to the table top or outside to a wall or fence.
- Fill the spray bottle with water.
- Squeeze 10-15 drops of food coloring into the spray bottle.
- Secure the top of the spray bottle.
- Prepare several spray bottles of water with different colors of food coloring.

Activity:

The child:

- Shakes the bottle.
- Stands next to the paper or sheet.
- Holding the spray bottle with one hand, sprays the colored water onto the paper.

The adult:

- Removes the tape and dries the mural.

Things To Think About

Basic Skill: Squeezing a trigger with one hand, using a palmer grasp.

Blowing Bubbles
Bubble Chase

Materials:

- Cups (plastic, paper, or styrofoam) half full of bubble solution
- Bubble solution of half water and half liquid dishwashing detergent
- Sugar
- Bubble blowing rings
- Open, outside area

Materials Preparation:

Mix bubble solution with two pinches of sugar per cup. This makes the bubbles last longer.

Activity:

The child:

- Dips the bubble blowing rings into the bubble solution.
- Blows on the ring to make bubbles (in the same direction as the wind).
- Chases the bubbles trying to recapture or break them.

The adult:

- Shows the children which direction the wind is coming from to blow the bubbles.

Things To Think About

Redirect the children to follow the wind so that the bubbles do not blow back into their faces.

Basic Skill: Running after and touching a moving target.

32

Going Fishing
Catch A Shape

Materials:

- Different shaped and colored sponges
- Scissors
- Water table, water tub, sink, bucket, etc.
- Buckets

Materials Preparation:

- Cut sponges into different shapes.
- Tape a different shaped sponge on the front of each bucket.
- Put the buckets next to the water table, bucket, sink, water tub, etc.
- Put the different shaped sponges in the water table, bucket, sink, water tub, etc.

Activity:

The child:

- Upon the adult's direction, picks up the specific shaped sponge.
- Squeezes the water out of the sponge fish.
- Puts the "fish" in the correct fish bucket.

The adult:

- Gives the child directions about which shape to pick up and put into the bucket:
  "Put the round fish sponge in the bucket,"
  "Put the square fish sponge into the bucket."

Things To Think About

Basic Skill: Squeezing sponges.

Sink or Swim
Boats That Float

Materials:

- A water table full of water
- Corks
- Rocks
- Sponges
- Bar soaps (including Ivory)
- Styrofoam pieces
- Toothpicks
- Short straws
- Paper
- 2 plastic buckets

Materials Preparation:

- Cut paper into the shapes of different types of sails.
- Put holes into paper for straws or toothpicks.

Activity:

The children:

- Test the various objects to see which float.
- Sort them into two buckets-floaters and nonfloaters.
- Select one floater.
- Construct a boat out of it by using a toothpick or straw as a mast and adding a paper sail.
- Blow the boats around the water table.

The adult:

- Shows each child how to push sail into the ''boat.''

Things To Think About

Basic Skill: Sorting into two categories.

34

Boats
Block Boats

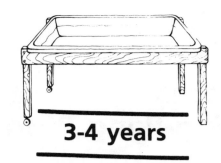

Materials:

- Water table, water tub, sink, bucket, etc.
- Clean styrofoam meat trays
- Straws
- Construction or typing paper

- Markers or crayons
- Scissors
- Pencils (unsharpened)
- Tape or glue
- One-hole punch

Materials Preparation:

- Cut out styrofoam squares in two sizes, small for boats and large for rafts.
- Outline a triangle on paper with marker.
- Mark two circles on the triangle to show the child where to punch holes with the hole punch.

Activity:

Styrofoam Block Boats and Rafts
The child:

- Cuts out a paper (construction paper or typing paper) triangle for a boat or square for a raft.
- Uses markers or crayons to decorate the sails.
- Tapes the paper triangle or square to the straw *or*
- Using the one-hole punch, punches a hole in the top and bottom (where marked) of the paper triangle or square.
- Pushes the straw through the paper sail.
- Pokes a hole in the styrofoam with the pencil.
- Places the straw sail into the hole in the styrofoam.
- Floats the styrofoam boats or rafts into the water.

The adult:

- Marks triangular shapes on construction paper for the child to cut out for sails.
- Washes the styrofoam meat tray.
- Outlines a triangle on paper with marker.
- Marks two circles on the triangle to show the child where to punch holes with the hole punch.

Things To Think About

Basic Skills: Pushing, cutting on a line, squeezing.

Races
Ping-Pong Push

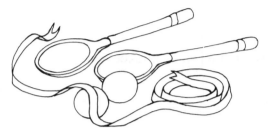

3-4 years

Materials:

- Large wooden spoons
- Ping-pong balls/plastic eggs
- Ribbon
- Tape

Materials Preparation:

None

Activity:

The children:

- Push ping-pong balls or plastic eggs in the water from one side of the water table to the other with a wooden spoon.

The adult:

- Shows each child how to push the ping-pong balls or plastic eggs to make them move through the water.
- Tapes or ties a ribbon down the middle of the water table, water tub, sink, bucket, etc.
- Has two children stand on either side of the ribbon.
- Clears the water table, water tub, etc. of all ping pong balls/plastic eggs but two.
- Puts one ping-pong ball or plastic egg on either side of the ribbon.

Things To Think About

Basic Skill: Palmer grasp, pushing an object, eye-hand coordination.

Water Has Energy!
My Water Wheel Turns

3-4 years

Materials:

- Water wheel
- Cups (plastic, paper, or styrofoam)
- Pots
- Pitchers
- Funnels
- Tubing
- Water table full of water

Materials Preparation:

None

Activity:

The children:

- Turn the water wheel by pouring water onto it from a variety of sources such as cups, pots, pitchers, etc.
- Notice differences in speed of water wheel and water sources.
- Collaborate to pour water through a variety of sources such as cups, tubing, pitchers and funnels before the water hits the water wheel.
- Note whether the water is faster when it travels through several sources.

The adult:

- Helps the children collaborate to make the water wheel turn.

Things To Think About

Basic Skill: Pouring at a target.

Sprinkler
Sprinkler Relay

Materials:

• Paper or styrofoam cups
• Sprinkler

Materials Preparation:

None

Activity:

The children:

• Divide up into relay teams.
• Run through the sprinkler, carrying a paper or styrofoam cup and catching water.
• A team member gives the cup filled with water to the next child in line.
• When the cup is filled with water, the children in that line sit down.

The adult:

• Helps children form teams.
• Puts children in two lines.
• Gives the children in the front of the line paper or styrofoam cups.

Things To Think About

Basic Skill: Running with an object in their hand.

Sprinkler
Having Fun with "Looby Loo"

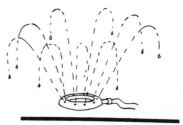

3-4 years

Materials:

• Sprinkler

Materials Preparation:

None

Activity:

The children:

• Form a circle around the sprinkler.
• Play "Looby Loo."
• Put named body part into sprinkler.

The adult:

• Leads the singing:
"Here we go looby loo.
Here we go looby li.
Here we go looby loo—
All on a Saturday night!

I put my right foot in.
I put my right foot out.
I give my foot a shake, shake, shake,
And turn myself about.''

Things To Think About

Basic Skill: Perform actions to a simple song.

Small Pool Blowing Bubbles

Materials:

- Small pool
- Straws and/or plastic tubing
- Goggles

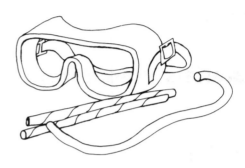

Materials Preparation:

None

Activity:

The children:

- Blow bubbles in the water with a straw/plastic tubing.
- Put their faces in the water and blow bubbles.

The adult:

- Shows each child how to blow bubbles in water with a straw.
- Shows each child how to put face in water and blow bubbles.

Goggles:

- If some children do not like their faces in the water, show them how to look at the water with goggles held to their faces but not strapped around their heads.
- After each child gets used to the goggles, place them on his/her head.
- Show each child how to clear the goggles if water gets in them.

Things To Think About

Basic Skill: Blowing.

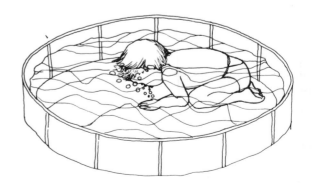

Small Pool
Follow the Leader

Materials:

• Small pool filled with water.

Materials Preparation:

None

Activity:

The children:

• Stand in the pool.
• Choose a leader.
• Follow the movements made by the leader.

The adult:

• Encourages the leader to sit in the water, put his/her face in the water and so on.

Things To Think About

Basic Skill: Wetting whole body.

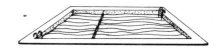

Large Pool
Let's Get into the Water

Materials:

• Large pool

Materials Preparation:

None

Activity:

The children:

• Sit on the edge of the shallow end of the pool.
• Sing ''This is the way I wet my toes'' to the tune of ''Here We Go 'Round the Mulberry Bush.''
• Continue with heels, ankles and calves as far as children can go.
  and
• Jump into the pool into an adult's arms.
• Sing ''This is the way I bubble, bubble.''
• Practice bubbling with face out of the water.
• Practice bubbling with face in the water.

The adult:

• Catches the child by the arms.

Things To Think About

Basic Skill: Bubbling in the water.

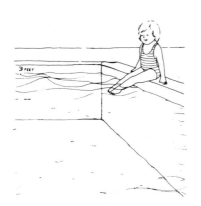

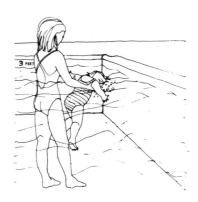

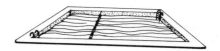

Large Pool Whirlpools

Materials:

- Paper cups
- Straws
- Large pool

Materials Preparation:

None

Activity:

The children:

- Sit on the side of the pool.
- Turn straws in a circular movement in cups of water, making the water swirl like a whirlpool. *and*
- Lift up arms and keeps legs straight while adult swings them in a circle in the water.

The adult:

- Gives each child a paper cup filled with water and a straw.
- Shows them how to turn the straw in a circle to make the water in the cup look like a whirlpool.
- Picks each child up under the arm and twirls the child in a circle in the water, creating a whirlpool action in the pool.

Things To Think About

Basic Skill: Contracting muscles in the body.

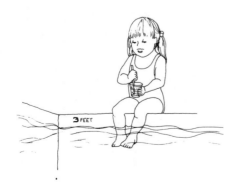

ACTIVITIES FOR
FIVE AND SIX YEAR OLDS

Pouring and Mixing
I Am a Measurer

Materials:

- Cups (plastic, paper, or styrofoam)
- Plastic pitchers
- Measuring cups
- Assorted colors of measuring caps (coffee and instant drink mix scoops)
- Water
- A shallow pan to catch spills if indoors
- Assorted colors of construction paper

Materials Preparation:

- Cut construction paper measuring caps out of colors which match the cups you have.

Activity:

The child:

- Fills a cup with a measuring cap.
- Takes a construction paper model of the measuring cap for each cap used.
- Counts how many of a particular cap it takes to fill the cup.
- Makes a pictograph of the measuring experience by placing the correct number of construction paper caps on a sheet of contrasting construction paper.
- Glues the construction paper measuring caps onto the construction paper.
- Repeats the activity with several different sizes and colors of measuring caps.

The adult:

- Helps child count the number of measuring caps used to fill the cup.
- Shows the child how to chart measuring caps on a pictograph.

Things To Think About

Basic Skill: Counting and graphing.

Pump Art
Chalk Designs

Materials:

- Colored chalk
- Plastic pump bottles in different sizes
- Water

Materials Preparation:

- Fill the plastic pump bottle with water.

Activity:

The child:

- Colors the sidewalk with colored chalk.
- Presses the pump top and lets the water splatter over the chalk on the sidewalk, making a water design on the chalk.
- Uses different sized pump bottles, comparing the different water marks (width, length, size, etc.) made by each bottle.

The adult:

- Shows the child how to press the pump to get the water out of the plastic bottle.

Things To Think About

Basic Skill: Pushing a pump top.

Squeezing Activities
Squeezing Science

Materials:

- Dishpan-⅔'s full of water
- Clear plastic squeeze bottles (large, medium, small, and clean)
- Basters
- Large, needleless syringes
- Assorted, waterproof, large squares, triangles and rectangles

Materials Preparation:

None

Activity:

The child:

- Selects a bottle with a lid.
- Squeezes the air out of it.
- Holds the tip underwater until the bottle stops filling.
- Holds the bottle upright.
- Squeezes more air out of it.
- Submerges the tip to continue filling it.
- Repeats the process until the bottle is filled.
  or
- Selects a waterproof shape.
- Puts it in the bottom of the dishpan.
- Pushes the plunger of the syringe down to the tip while the tip is in the water.
- Observes the air bubbles coming out of the syringe.
- Pulls the plunger back to fill the syringe with water.
- Squeezes the water out of the syringe while tracing around the shape in the bottom of the pan.

The adult:

- Shows the child that a VACUUM is created when the air is emptied out of the bottle and water rushes in to fill the vacuum.
- Points out with the child that if the tip of the bottle is not held underwater, air rushes in to f the vacuum.
- Points out that if you hold your finger over the tip of the syringe when the plunger is down to the tip, it is extremely difficult to pull it back.

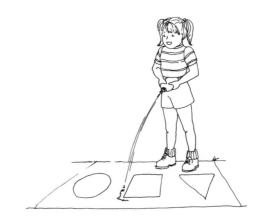

Things To Think About

This activity can be done outside with chalk shapes drawn on the ground.

Basic Skill: Traces around a shape.

Spray Art
Designer T-Shirt

Materials:

- Spray mist bottle with trigger handle
- Food coloring
- Large white butcher block paper
- Scissors
- Tape
- Large towel or plastic mat
- Water
- T-shirt

Materials Preparation:

- Spread towel or plastic mat outside, on the floor, or under a table.
- Place another towel or mat on the table.
- Tape a white T-shirt to the towel, mat, or to the towel on the table.
- Fill the plastic squeeze spray bottle with water.
- Squeeze 10-15 drops of food coloring into the plastic spray bottle.
- Secure the top of the spray bottle.
- Prepare several plastic spray bottles of water with different colors of food coloring.
- Shake the bottles.

Activity:

The child:

- Shakes the plastic spray bottle.
- Stands next to the table, towel, or mat.
- Holding the spray gun bottle in one hand, sprays the colored water onto the T-shirt.

The adult:

- Removes the tape and hangs the T-shirt to dry.

Things To Think About

Add more food coloring if color is too light. If you want to be sure of permanent color, use liquid dye instead of food coloring and water.

Basic Skill: Squeezing a trigger handle with one hand.

49

Blowing Bubbles
Bubble Sculptures

Materials:

- Unbreakable cups half full of bubble solution
- Bubble solution of half water and half liquid dishwashing detergent
- Sugar
- Bubble blowing rings
- Slotted spoons (the type for serving vegetables)

Materials Preparation:

- Mix bubble solution with two pinches of sugar per cup. This makes the bubbles last longer.

Activity:

The child:

- Dips a bubble blowing ring into the bubble solution.
- Blows a bubble and keeps it on the ring.
- Observes how the light reflects off the bubble.
- Observes how the bubble changes shape as the ring is moved.
  and
- Repeats the activity with a slotted spoon and the multiple bubble sculpture that comes from blowing through the slotted spoon.

The adult:

- Demonstrates making a sculpture with a slotted spoon.

Things To Think About

Basic Skill: Controlling soap bubbles.

Going Fishing
Catching Words and Numbers

5-6 years

Materials:

- Colored sponges
- Scissors
- Water table, water tub, sink, bucket, etc.
- Numbered or labeled bucket
- Markers
- Paper
- Clear contact paper or laminating machine
- Tape or glue
- Buckets

Materials Preparation:

- Cut sponges in the shape of fish.
- Write numbers or words on paper.
- Laminate or put clear contact paper on the number or word.
- Cut out the words or numbers.
- Tape the numbers or words to the sponge fish.
- Glue different numbers or words on the sponge fish.
- Tape a sample fish (specific number or word) onto the front of a bucket.
- Put the buckets next to the water table, bucket, sink, water tub, etc.
- Put the sponges in the water table, bucket, sink, water tub, etc.

Activity:

The child:

- Picks up the fish with the specific number or word that is called. "Pick up all the fish with numeral 4." or "Pick up all the fish with the word cat."
- Squeezes the water out of the sponge.
- Puts the "fish" in the fish bucket with the same numeral or word taped on it.

The adult:

- Gives verbal direction to the child, "Pick up the fish with the numeral 4." or "Pick up the fish with the word cat."

Things To Think About

Basic Skills: Squeeze sponges, matches and identifies a specific numeral or word.

Sink or Swim
Can You Make It Sink?

5-6 years

Materials:

- Watertight film canisters
- Dried beans
- Styrofoam pellets
- Small pebbles
- Dried peas
- Posterboard
- Markers

Materials Preparation:

- Test all of the objects in canisters to determine the largest number it will take to make the canister sink or filled with the objects.
- Make a grid on a piece of posterboard large enough to hold the maximum number of objects it can take to make a canister sink or filled.
- Include a column on the grid for a sink or float symbol.
- Make sink symbols shaped like anchors and float symbols shaped like boats.

Activity:

The children:

- Place a bean, pea, pellet, or pebble in a canister.
- Seal the canister.
- Test to see if it will sink.
- Repeat until the canister sinks or is full.
  and
- Make a graph of their results by emptying the canisters and laying the contents on the posterboard grid.
- Indicates whether the canister sank or floated by placing an anchor or boat symbol in the sink or float column of the grid.

The adult:

- Assists with opening and sealing the film canisters.
- Helps each child make a graph.

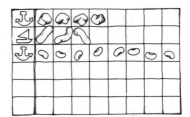

Things To Think About

The children may need help sealing and opening the film canisters.

Basic Skill: Guessing an outcome to a problem. Experimenting until a goal is reached.

Boats
Soap Carving

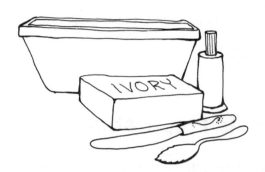

5-6 years

Materials:

- Water table, water tub, sink, bucket, etc.
- Bar of Ivory soap
- Butter knife
- Grapefruit spoon
- Markers

Materials Preparation:

- Outline the area of the soap that the child wants to carve and scoop.

Activity:

The child:

- Carves the soap into a boat shape, following the marker lines.
- Scoops out the top of the soap (where marked) to make the boat seat.
- Floats the soap boat in the water.

The adult:

- Shows the child how to shave the soap with a butter knife.

  **Shows the child safety cutting tips:
  Always begin the shaving process,
  pointing the blade of the knife to the
  outside.
  Make the shaving stroke beginning from
  the inside of the body to the outside.**

- Shows the child how to scoop out the piece of soap with a grapefruit spoon.

Things To Think About

Basic Skill: Carving, scooping.

Races
Ping-Pong Blow

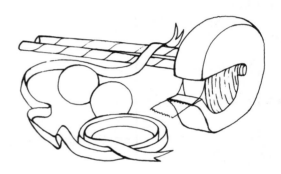

5-6 years

Materials:

- Water table, water tub, sink, bucket, etc.
- Straws
- Ping-pong balls or empty plastic eggs
- Water
- Ribbon
- Tape

Materials Preparation:

- Put ping-pong balls/plastic eggs in the water table, water tub, sink, bucket, etc.

Activity:

The children:

- Blow air through straws to make the ping-pong balls/plastic eggs move through the water.
- Race with two ping-pong balls/plastic eggs.

The adult:

- Shows each child how to blow air through the straw to make the ping-pong balls or plastic eggs move through the water.
- Tapes or ties a ribbon down the middle of the water table, water tub, sink, bucket, etc.
- Has a child stand on either side of the ribbon.
- Clears the water table, water tub, etc. of all ping pong balls/plastic eggs but two.
- Puts one ping-pong ball or plastic egg on either side of the ribbon.

Things To Think About

Basic Skill: Blowing through a straw.

Water Has Energy
A Rube Goldberg!*

5-6 years

Materials:

- Styrofoam cups
- Pencils
- Funnels
- Plastic tubing
- Strainers
- Water wheel(s)
- A water table full of water
- Pegboard with wires through it and hooks on it

Materials Preparation:

None

Activity:

The children:

- Poke holes in styrofoam cups with the pencils.
- Tie or hang their cups, funnels, tubing and and strainers to the pegboard to create a continuous flow when water is poured in at the top.
- Hold "Rube Goldberg" water wheel turner above the water wheel in the water table.
- Pour water in at the top.
- Watch it pour through the system and turn the water wheel.
- Remember water has energy.
- Redesign the system again.

The adult:

- Supervises carefully.
- Helps children tie securely.
- Helps children troubleshoot the design if it doesn't work.

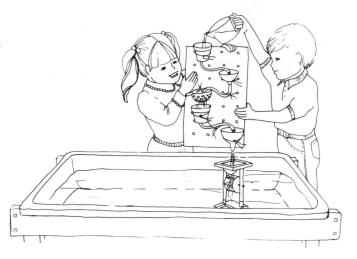

Things To Think About

Remind the children that water turns the wheel because it has energy.

Basic Skill: Problem solving: Planning a system.

* Rube Goldberg was a cartoonist who drew complicated inventions to do simple jobs.

55

Sprinkler
Fill and Measure

Materials:

- Cylinders
- Plastic or styrofoam cups
- Different shaped plastic containers (round, square, retangular, etc.)
- Plastic measuring cup
- Plastic bowls (same size)
- Paper
- Markers

Materials Preparation:

None

Activity:

The children:

- Hold paper or styrofoam cup, cylinder, or plastic container.
- Run through the sprinkler and fills the cup, cylinder, or plastic container up with water.
- Line up the filled different containers.
- Pour water from filled container into a measuring cup.
- Measure the amount of water in each container.
- Pour the water into plastic bowls.
- Compare which containers hold the same amount of water, which containers hold more, which containers hold less, etc.

The adult:

- Give the children different plastic, paper, or styrofoam containers to be filled with water.

Things To Think About

Basic Skill: Pouring; running; measuring; comparing.

Sprinkler
Making Rainbows

Materials:

• Prisms
• Sprinkler
• Sunny day

Materials Preparation:

None

Activity:

The children:

• Hold prisms in the sprinkler.
• Make as many colors as possible.

The adult:

• Helps each child see a color.
• Helps each child hold the prism.

Things To Think About

Basic Skill: Recognize a rainbow.

Small Pool
Sand Sculpture

Materials:

- Small pool
- Bucket
- Sand
- Waxed paper
- 3 cups of water
- Tape

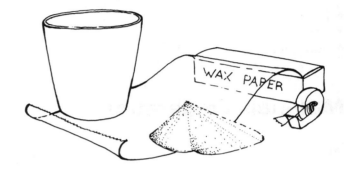

Materials Preparation:

Tape large waxed paper squares next to the small pool.

Activity:

The children:

- Take hands full of wet sand out of the bucket.
- Drip the sand onto pieces of waxed paper next to the pool, making sand sculpture structures.

The adult:

- Scoops 2 cups of sand in a bucket.
- Puts the bucket filled with sand and water next to the pool.
- Mixes 3 cups of water with the sand so that the sand is a very wet and runny consistancy.
- Shows each child how to take a hand full of sand and drip the sand onto the waxed paper to make a dripped structure.

Things To Think About

Basic Skills: Scoop, pincer grasp, release.

Small Pool Simon Says

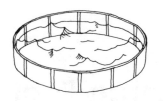

Materials:

• Small pool filled with water

Materials Preparation:

None

Activity:

The children:

• Stand in the pool.
• Choose a Simon.
• Follow the instructions given by Simon provided that "Simon Says."

The adult:

• Demonstrates how to play "Simon Says" by being the first leader.

Things To Think About

Basic Skill: Participate in a game.

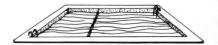

Large Pool
Kickboards

Materials:

• Kickboards

Materials Preparation:

None

Activity:

The children:

• Hold onto styrofoam kickboards at the end of the board and kick in the water, moving through the water.
  or
• Put stomachs on kickboards, holding the board near the top, kicking and moving through the water.

The adult:

• Gives each child a kickboard.
• Walks alongside the children as he/she moves on the kickboard through the water.

Things To Think About

Basic Skill: Kicking.

Large Pool
Doggie Paddle

Materials:

• Large pool

Materials Preparation:

None

Activity:

The children:

• Sing "I'm a Little Swimmer" to the tune of "I'm a Little Teapot."
• Practice the paddle and kick outside of the water.
• Get into the water and dog paddle while the adult sings.

The adult:

• Teaches the song to the children:
    I'm a little swimmer.
    Look at me
    Paddle my arms
    And kick my legs.

    When I get all steamed up,
    Then I shout:
    "Watch me swim
    And please look out."

Things To Think About

Basic Skill: Practice swimming strokes.

BIBLIOGRAPHY

Bayley, N. (1969). Bayley Scales of Infant Development. New York: Psychological Corporation.

Bluma, M., Shearer, A., & Frohman, J. (1976). Portage Guide to Early Education. Portage, Wisconsin: Cooperative Educational Service.

Bower, T.G.R. (1974). Development in Infancy. San Francisco: W.H. Freeman & Co.

Brigance, Albert. Brigance Diagnostic Inventory of Early Development. Massachusetts: Curriculum Associates.

Broch, Kenneth & League, R. (1970). Receptive Expressive Emerging Language Skills. Baltimore: University Press.

Chase, Richard A. & Rubin, Richard R., ed. (1979). The First Wondrous Year. New York: Collier Books.

Doll, E.A. (1966). Preschool Attainment Record. Circle Pines, Minnesota: American Guidance Service.

Doll, E.A. (1986). Vineland Adaptive Behavior Scale. Circle Pines, Minnesota: American Guidance Service.

Gesell, A. (1940). The First Five Years of Life. New York: Harper & Row.

Glover, Elayne, Preminger, J. & Sanford, A. (1978), The Early Language Accomplishment Profile for Developmentally Young Children. Winston-Salem, North Carolina: Kaplan Press.

Hendrick, Donna, Prather, E., & Tobin, A. (1975). Sequential Inventory of Communication Development. Seattle, Washington: Univ. of Wash. Press.

Kiernan, Sharon & Conner, F. (1978). Chart of Normal Development. Washington: Government Printing Office.

Koch, Jaroslav. (1976). Total Baby Development. New York: Pocket Books.

Leach, Penelope. (1978). Your Baby & Child. New York: Alfred A. Knopf.

Levy, Janine. (1975). The Baby Exercise Book. New York: Random House.

Montgomery, Patricia & Richter, Eileen. (1973). Sensory Integration Program Checklists. Los Angeles: Western Psychological Services.

Painter, Genevieve. (1971). Teach Your Baby. New York: Simon & Schuster.

Rubin, Richard R., Fisher, John J. III, Doering, Susan G. Your Toddler. (1980). New York: Collier Books.

Sanford, Anne & Zelman, Jan. (1981). The Learning Accomplishment Profile (Rev. Ed.). Winston-Salem, North Carolina: Kaplan Press.

Santa Cruz County Office of Education. (1973). Behavioral Characteristics Progression. Palo Alto, California: VORT Corporation.

Zimmerman, I., Stainer, V. & Evatt, R. (1969). Preschool Language Scale. Columbus, Ohio: Charles E. Merrill.